IMAGES
of America

LAKE MICHIGAN'S AIRCRAFT CARRIERS

IMAGES
of America

LAKE MICHIGAN'S AIRCRAFT CARRIERS

Paul M. Somers

ARCADIA
PUBLISHING

ISBN 978-0-7385-3208-0

Published by Arcadia Publishing
Charleston SC, Chicago IL, Portsmouth NH, San Francisco CA

Printed in the United States of America

Library of Congress Catalog Card Number: 2003112722

For all general information contact Arcadia Publishing at:
Telephone 843-853-2070
Fax 843-853-0044
E-mail sales@arcadiapublishing.com
For customer service and orders:
Toll-Free 1-888-313-2665

Visit us on the Internet at www.arcadiapublishing.com

Contents

ACKNOWLEDGMENTS

Baha'i House of Worship
Emil Buehler Library
Buffalo Police Department
Chicago Maritime Society
Cleveland Public Library
Dossin Great Lakes Museum
Frederick Clark Durant, III
Great Lakes Historical Society
Milwaukee Public Library
National Archives
National Baha'i Archives
National Museum of Naval Aviation
San Diego Maritime Museum
US Naval Historical Center

INTRODUCTION

Veterans, historians, and authors often express opinions as to what person, what weapon, or what piece of equipment "won World War II." The one thing which all such pronouncements have in common is that they are all wrong.

No person—neither General Patton nor Audie Murphy, and no weapon—neither the A-bomb nor the M-1 rifle, and no piece of equipment—neither the *USS Missouri* nor the Willys Jeep single-handedly brought an end to the conflict. Substantial contributions to victory came from many sources. This is the story of two ships which played a major role in the defeat of the Japanese without ever leaving United States territorial waters: The *USS Wolverine* and the *USS Sable*.

From the first moment of the Japanese sneak attack on Pearl Harbor, it was evident that naval warfare in World War II would not consist of Jutland-type engagements with battleships hurling projectiles at one another within clearly-defined, relatively small areas. Though it was published a few months too late to save the Pacific Fleet, one of the most comprehensive analyses of what had happened to naval warfare and what was going to happen appeared, unlikely enough, in the July 1942 issue of *National Geographic Magazine* in an article by Melville Bell Grosvenor entitled "The New Queen of the Seas."

In that article, Grosvenor argued the need for more floating airbases—flight decks on cargo ships, on troop carriers, on tankers; indeed, on almost every type and size of ship involved in the war effort. His thesis was that such ships could deliver assembled and manned airplanes within flying distance of where they were needed, launch them, and see them delivered, ready for action, not crated, unassembled and with no simultaneously-delivered pilots! Such ships could also be defended en route by their cargo—launching their airplanes to meet an enemy attack by surface or air.

Only on the next to last page of this article is there even the slightest mention of an adequate supply of pilots, and then, only obliquely in the caption of a photograph. It was under a photo of the Great Lakes cruise ship *SS Seeandbee* being converted to a practice carrier to train young flyers in take-offs and landings in preparation for combat situations.

This is a story of how the *SS Seeandbee* and another Great Lakes cruise ship, the *SS Greater Buffalo*, made a major contribution to the defeat of the Japanese in World War II.

One

Before Pearl Harbor

The concept of ship-borne aircraft in the U.S. Navy began on November 14, 1910, when Eugene Ely took off from an 83-foot downward-slanted flight deck which had been built on the bow of the light cruiser *USS Birmingham*. He flew his 50 horsepower Curtiss biplane to a safe landing on land. This was the first ever aircraft take-off from a ship.

On January 18, 1911, Mr. Ely, again, made history by landing on a ship. This was done on a 102-foot platform added to the stern of the armored cruiser *USS Pennsylvania*. The deck was fitted with a simple arrestor system of cables, weighted down by sandbags, which was designed to catch a hook attached to the bottom of Ely's plane.

Both of these exercises were considered demonstration; it took until 1922 for the U.S. Navy to acquire its first dedicated carrier, the *USS Langley* (CV1). It was a conversion of the collier *USS Jupiter*. The *Langley* was 542 feet long with a displacement of 12,000 tons and a top speed of 14 knots. Its normal complement rose from 8 planes to 30 during its career. This flush-decked carrier served until 1936 when the forward one-third of its flight deck was removed in its refitting and reclassification as a seaplane tender (AV3). It was sunk by the Japanese early in World War II while ferrying Army Air Corps P-40s to Java.

As the U.S. Navy's only commissioned aircraft carrier from 1922 until the completion of the *USS Lexington* (CV2) and the *USS Saratoga* (CV3) in 1927, the *Langley* served as the principal laboratory for the development of early naval air tactics.

By the day before the Pearl Harbor attack, the Navy had seven commissioned aircraft carriers: the *Lexington* (CV2) and the *Saratoga* (CV3), which had both been laid down as cruisers; the *USS Ranger* (CV4), whose design, limited by the terms of the Washington Treaty, was considered ill suited for service in the Pacific, so it spent all of World War II in the Atlantic; the *USS Yorktown* (CV5), the *USS Enterprise* (CV6), and the *USS Hornet* (CV8); and the *USS Wasp* (CV7), the last of the carriers built under the Washington Treaty, shorter than the *Ranger* and with no armor.

On the day of the Japanese attack on Pearl Harbor, four of the Navy's carriers were in the Atlantic; *Hornet*, *Ranger*, *Wasp*, and *Yorktown*. Fortunately, the three carriers stationed in the Pacific—*Lexington*, *Saratoga*, and *Enterprise*—were away from Pearl Harbor, ferrying Marine planes to Pacific island outposts.

By October 1942, less than 11 months into World War II, the *Lexington*, *Yorktown*, *Hornet*, and *Wasp* had all been sunk. The *Saratoga* had been severely damaged in January of 1942, repaired, and returned to action in June. Torpedoed again in August, it returned to San Diego. The *Enterprise* was damaged and returned to Pearl Harbor in August.

These early engagements and losses emphasized that carriers would be the backbone of war in the Pacific. While American industry would build carriers and their airplanes, it was up to the Navy to train the pilots and crews that would be necessary to make the ships and planes effective.

In the fall of 1941, months before the attack on Pearl Harbor, (then Commander, later Vice Admiral) R.F. Whitehead, aviation aide to the commandant, Ninth Naval District (Great Lakes) originated the idea of training carrier pilots on the Great Lakes.

With war seemingly approaching and German submarines active close to the east coast, the Atlantic would have been a poor choice as a training site. When the United States was attacked in Hawaii and so much of the Pacific Fleet had been destroyed or disabled, it was widely feared that Japanese naval units would be operating just off the Pacific coast. Also, among civilians, at least, there was a genuine expectation of a Japanese invasion of the west coast.

Commander Whitehead's suggestion of Lake Michigan, a body of water totally within the United States, meant that a training carrier would not need armor or armament, escorting destroyers, as many radio restrictions, and could be supported out of Chicago, which had significant naval presence, including the Great Lakes Naval Training Center, Naval Air Station, Glenview, and Naval Aviation Mechanics' School on Navy Pier.

Commander Whitehead's plan prompted a letter from Rear Admiral John Downes, Commandant, Ninth Naval District, to the Chief of Naval Operations, which said, in part:

Such operations could be conducted in waters protected from mines and submarines. When not otherwise employed, the carrier could be moored to the Navy Pier at Chicago where the largest naval aviation mechanics' school in the United States is located.

Also, Lake Michigan's cold winters would allow the Navy to experiment with foul weather gear. As it turned out, operations were interrupted on Lake Michigan only during the winter of 1942–43.

Richard Francis Whitehead who, as aviation aide to the Commandant, Ninth Naval District, originated the idea of training carrier pilots on the Great Lakes. (Courtesy of U.S. Naval Historical Society.)

Eugene Ely's Curtiss pusher aircraft being hoisted aboard the *USS Birmingham* preparatory to the first take-off from a U.S. Navy ship. (Courtesy of U.S. Naval Historical Society.)

Eugene Ely's Curtiss pusher aircraft aboard the *USS Birmingham* ready to make the first take-off from a U.S. Navy ship. (Courtesy of U.S. Naval Historical Society.)

Eugene Ely is seen here landing his Curtiss pusher aircraft on the *USS Pennsylvania.* (Courtesy of U.S. Naval Historical Society.)

Eugene Ely in his Curtiss pusher aircraft returning to shore from a successful landing on the *USS Pennsylvania*; note sandbagged arrestor wires. (Courtesy of U.S. Naval Historical Society.)

Collier *USS Jupiter* is seen here before its conversion to the *USS Langley*, the U.S. Navy's first aircraft carrier. (Courtesy of U.S. Naval Historical Society.)

The *USS Langley* cruising off San Diego with the *USS Somers* (no relation to the author) in the background. (Courtesy of U.S. Naval Historical Society.)

The *USS Langley* after conversion to a seaplane tender; note shortened flight deck. (Courtesy of U.S. Naval Historical Society.)

The *USS Lexington* is seen launching Martin T4M torpedo bombers *c.* 1929. (Courtesy of Miles Desomer.)

Two

SS SEEANDBEE

Owner: Cleveland and Buffalo Transit Company
Builder: Detroit Shipbuilding Company, Wyandotte, Michigan, 1913
Dimensions: 500 feet x 58 feet x 23 feet 9 inches

A wise Greek philosopher observed that two abilities were necessary for success in any venture: the ability to see with your eyes closed and the courage to act on what you saw. Commander Whitehead exhibited both that ability and that courage.

Once the U.S. Navy had agreed with his choice of Lake Michigan as the location for the carrier pilot qualification training program, the next priority was to secure an aircraft carrier.

Using a fleet carrier was out of the question for two reasons; by the time the decision had to be made to adopt the program, the United States was at war and needed all its commissioned carriers at sea. Further, no existing carrier was narrow enough to use the Welland Canal to sail from the Atlantic Ocean to the Great Lakes. The same Welland Canal width restriction barred the conversion and use of other ocean going ships, with otherwise appropriate specifications, as well. (Think how proud Chicagoans would have been to have a re-floated and refitted SS *Normandie* sailing just offshore!)

This left the Navy with two choices: build a new aircraft carrier from scratch or convert an existing ship to an aircraft carrier. Building a carrier would have taken too long and would have tied up shipbuilding capacity needed for other war-related construction. Thus, conversion of a ship already on the Great Lakes was determined to be the way to go.

The Auxiliary Vessels Board recommended that the training carrier should have a 500-foot flight deck and be capable of a speed of 18 knots.

In Commander Whitehead's survey of available ships, the choice for conversion narrowed down to two vessels: the *SS City of Midland*, a car ferry owned by the Pere Marquette Railroad, and the *SS Seeandbee*, a Great Lakes excursion cruiser. Because the car ferry was shorter, slower, and already making a contribution to the war effort, which the *Seeandbee*, a recreational ship, was not, the Seeandbee became the vessel of choice.

The *Seeandbee* had been built in 1913 to cruise Lakes Erie, Huron, and Michigan. The Soo Locks and the Welland Canal were too narrow for it to sail Lakes Ontario and Superior. Its name had been chosen in a contest sponsored by its owners and won by May Knight of Cleveland. The name was based upon the corporate name of the owner, the Cleveland & Buffalo Transit Company. Miss Knight won $ 10 and a free trip on the ship for her suggestion!

The *Seeandbee* had a dark green hull, white upper decks, and four black stacks. At 500 feet, the *Seeandbee* was long enough to accommodate a usable flight deck; it was fast enough, with a modest headwind, for adequate take off and landing speeds and, being a side wheeler, it had great resistance to rolling. Its main disadvantages were that it was coal-fired at a time when the rest of the Navy was oil fired, and it had a tall superstructure filled with very ornate and expensive fittings which would have to be removed and scrapped.

When Commander Whitehead evaluated the *Seeandbee*, he found that it was not, in its current use and configuration, an essential factor in pursuing the war, and, therefore, deserved further consideration. His observations of the ship included the following:

She had been launched on November 9, 1912, and was new enough to be modern, but old enough to be somewhat obsolete even in its then-current use. In 1912, it had been the largest side-wheeled passenger ship in the world. Architect Frank E. Kirby said, at the time, he would never design a bigger side-wheeler. Several years later, this turned out to be an inaccurate prediction.

Her guaranteed speed from dock to dock was 22 miles per hour. With a modest headwind, this would be more than enough for aircraft take-offs and landings.

It had 24 parlors with private baths and toilets, 62 staterooms with private toilet facilities, and 408 regulation staterooms. Design of the public facilities was the work of marine interior decorator Louis O. Keil. Simplicity was the hallmark of Keil's design.

Passengers entered a lobby as large as that of a fine hotel. It included the steward's office, the purser's office, telephone booths, baggage rooms, and a lunch counter. The walls were paneled with mahogany inlaid with other woods. The ceiling light fixtures were bronze. At one end was the grand stairway leading to the promenade deck. This stairway was enclosed in a vestibule with sliding doors, which would shut off the stairway in the remote event of a fire.

Aft of the lobby, on the main deck, was the main dining room, 72 feet long by 60 feet wide forward tapering to 32 feet aft, which seated 170. On the starboard side was a 24-foot banquet room and on the port side, two private dining rooms. There were a number of alcoves on both the port and starboard sides which afforded guests a measure of privacy while dining. Light was provided by Sheffield silver candelabras and wall fixtures. Aft in the main dining room was a great sideboard with a dumb waiter connected to the buffet on the orlop deck below.

Locating the dining room on the main deck, a feature pioneered on the *SS City of Detroit II*, permitted the diners to look out over the water as the ship sped along.

The buffet, on the orlop deck directly below the dining room, was reached by a stairway from the main deck. Its design was that of an old English tavern. It inspired *The Marine Review* to cite Samuel Johnson's observation that nothing was ever contrived that gave mankind so much creature comfort as a good tavern.

The 400-foot-long main saloon on the promenade deck was reached via the grand stairway from the lobby. Its features included flower booths, a book and periodical store, an observation room, and men's and women's writing rooms. The rooms were finished in mahogany wainscoting below and enamel above.

The color scheme of the gallery deck was gray, ivory, and white. The aft end of this deck was a ladies' drawing room of Italian renaissance design. Furnishings for this room were walnut covered with tapestry and Wilton floor covering. The color scheme was rose.

The atrium was located on the upper deck immediately above the drawing room. It was designed to resemble a Pompeian court with adjoining sleeping rooms. Lighting was by bronze torches and the ceiling was painted to represent the sky. In the center of this room was an open well looking down upon the drawing room.

The balcony for the orchestra was located at the after end of the main saloon just forward of the ladies' drawing room on the gallery deck above. This placement of the orchestra allowed its music to reach, not only the main saloon, but the drawing room and the atrium above.

Amidships on the gallery deck was the lounge. This room could be reached by a forward stairway from the promenade deck between the stacks to the upper deck. The stairway was lighted with the L'Art Nouveau electric light fixtures on the newel posts. The room was finished in fumed oak with L'Art Nouveau decoration painted directly on the wood. There were numerous bays, both port and starboard, where light refreshments could be enjoyed.

Surrounding the main saloon were a series of parlors named in honor of the president, general manager, traffic manager, and directors of the line. Each parlor was of an individual design finished in vermillion wood, satin wood, mahogany, red gum, silver gray maple, primavera, or enamel. The furnishings and decoration in each parlor matched the character of the wood finish and included two brass beds, a divan, tables, dressers, chairs, mirrors, and cushions. Each had a private bath finished in white enamel, and each had a private balcony. The lighting fixtures were Sheffield silver and Tuscan gold.

Second class cabins were all steel and were located forward of the lobby on the main deck.

There were fourteen 40 person lifeboats, two 20 person lifeboats, and two 16 person lifeboats. There were also life rafts and life preservers as required by the United States Steamboat Inspection Service.

Ventilation was provided by a washed air system serving all inside staterooms, dining room, buffet, smokers' room, galley, crew's quarters, lavatories, and toilet rooms. Air in the staterooms entered under the lower berth so that no draft is felt in the room. Vents from all toilet fixtures were connected by ductwork to aspirating tubes in two of the stacks, maintaining positive exhaust making it impossible for odors to escape into the ship.

In addition to making the ship a popular cruise ship, all of these luxurious features caused the Republican National Committee to press the *Seeandbee* into service as a floating hotel for delegates to its 1936 National Convention, held in Cleveland, which nominated Kansas Governor Alf Landon to run against Franklin D. Roosevelt.

Commander Whitehead was not put off by all this grandeur, which would have to be scrapped to make way for a flight deck and bridge. He saw that there were many features of the *Seeandbee* that were appropriate for the conversion.

There was an all-steel hull with a 365-foot double bottom for water ballast. This double bottom was divided at the centerline by fore and aft watertight girders and was further subdivided into 14 separate watertight compartments.

Above the double bottom, the ship was divided into 11 watertight compartments with bulkheads extending from the keel to the main deck. Bulkheads, which were permitted to have doors, were fitted with watertight doors operated hydraulically from the engine room.

The extensive use of steel rendered that section of the ship practically fireproof. The crew's quarters were built of steel and the cargo spaces were insulated with galvanized iron and asbestos wherever wood was used.

Fire hydrants were located throughout the ship and spaced so that 50-foot lengths of hose could reach every part of the ship. An automatic sprinkler system covered every part of the interior of the ship. The system was controlled by a sprinkler pump located in the engine room.

Two trimming tanks, each with 52 tons of capacity were located on the port and starboard sides just aft of the wheel casings. They could be filled or emptied in less than four minutes, making it possible to keep the ship on an even keel.

Quick handling in tight places was insured by a bow rudder, necessitated by the tricky channels at both Buffalo and Cleveland. There was, of course, also a stern rudder. Both rudders were steam operated.

The ship was fitted with two 6,500 pound anchors each with 185 fathoms of 2 1/4 inch steel chain cable.

The ship's power plant was designed to produce 12,000 horsepower. There were nine boilers in the steam plant: six of single end design (14 feet in diameter and almost 11 feet long) and three of double end design (14 feet in diameter and almost 22 feet in length) all built for a working pressure of 165 pounds. There were four jacketed smokestacks separated from the cabins with insulated steel casings.

The main engine was located in a separate compartment aft of the boilers. The engine was a three-cylinder, compound, inclined type with on 66 inch diameter high pressure cylinder and two 96 inch low pressure cylinders. All three had a stroke of 108 inches. They were the largest engine cylinders ever cast in the Great Lakes region.

The two paddle wheels were 30 feet in diameter over the outer edge of the buckets and 32 3/4 feet in diameter over the rims. There were 11 buckets on each wheel, 14 feet 10 inches long, 5 feet 1 inch wide and of 1 3/8 inch steel plate. The wheels weighed about 100 tons each and were built to deal with the severe ice conditions at Buffalo in the early spring.

The 4500 lights and other electrical gear on board were powered by three 75 kilowatt turbine-driven generators. The main switchboard was in the engine room on the main deck. The circuits were designed so that an electrician could disconnect any combination of lights, leaving light enough to get around. This was especially convenient for night lighting the ship.

There were over 500 telephones on board: one in every stateroom, in the officers' quarters, and in the telephone booths. There was also a private telephone system for the operation of the ship, from the pilot house to the engine room and other operating locations.

Signal lights were carried in duplicate. If the primary light burned out, an auxiliary light cut in and a signal was sent to the pilot house.

There was an elaborate system of fans throughout the ship, especially in the dining room. Drinking water was purified by electricity before being served. Many of the functions in the galley were performed by electricity—even potato peeling and dishwashing (and this in 1914!).

The ship was equipped with a 32 inch searchlight, the largest on the Great Lakes. The forward funnel carried one 10 inch and one 26 inch whistle and a 6 inch organ whistle.

Ashes are discharged outboard by eight hydraulic double-jet ash ejectors. The stoke holds were large and well-ventilated and were designed with safety escapes and for the comfort of the stokers.

It can be seen that, though 30 years old, the ship had many operating features which commended it to Commander Whitehead for conversion, but the beauty and ornamentation that contributed to making it a popular cruise ship would have to be demolished.

After convertibility, economics was the next consideration.

The *Seeandbee* had been built in 1913 at a cost of $1,600,000. In 1939, the destruction by fire of the *SS City of Buffalo* in 1938 coupled with the Great Depression and increasing competition from trucks and railroads brought about the bankruptcy of the *Seeandbee's* owner, The Cleveland & Buffalo Transit Co. As part of the liquidation, the *Seeandbee* was sold for $135,000 to T.J. McGuire whose newly-organized C&B Transit Co. of Chicago operated it on a regular schedule on Lake Erie through 1941.

Commander Whitehead had Rear Admiral W.A. Lee, Jr., head of fleet training division, send two officers, one from the bureau of ships and one from Fleet Training Division., to inspect the vessel. They agreed with Commander Whitehead's opinion. They were especially impressed by the width over the sponson deck, which, at 98 feet, was wider than the deck of the *USS Essex*.

C&B Transit Company, reflecting upon frozen prices, lack of replacement parts, the shortage of manpower, and a $756,500 offer, agreed to sell the *Seeandbee* to the U.S. Navy on March 10, 1942.

SS City of Midland, a car ferry owned by the Pere Marquette Railroad, was the runner up in the search for an appropriate ship for conversion to the first aircraft carrier on the Great Lakes. (Courtesy of Great Lakes Historical Society.)

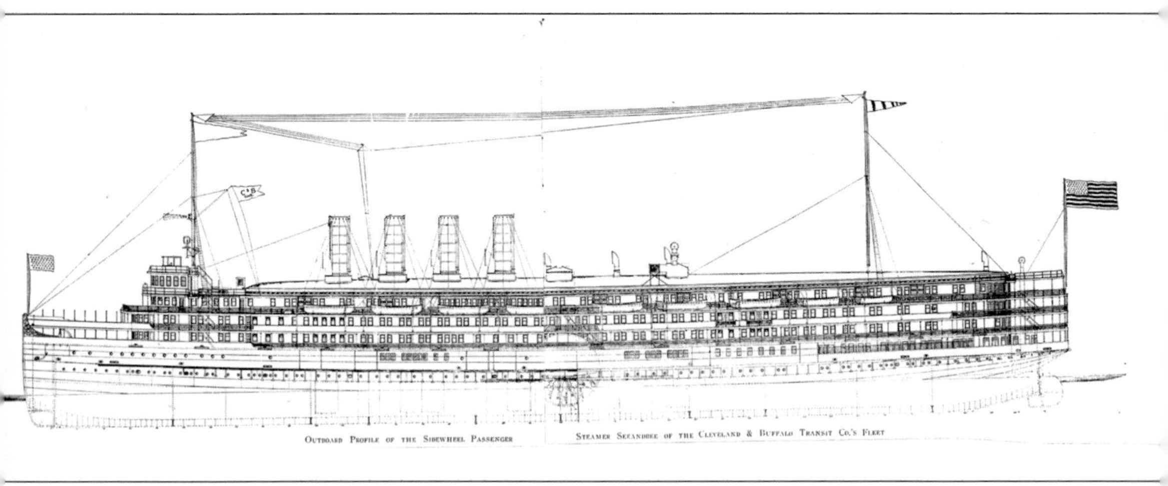

This is the outboard profile of the *SS Seeandbee*. (Courtesy of *Marine Review*.)

SS *Seeandbee*, shown in port and ready for a Great Lakes cruise, was chosen as the first Great Lakes ship to be converted to an aircraft carrier. (Courtesy of Cleveland Public Library.)

SS Seeandbee is seen here in port and ready for passengers. (Courtesy of Great Lakes Historical Society.)

SS Seeandbee ticket office is shown ready for passengers. (Courtesy of Great Lakes Historical Society.)

SS Seeandbee is loaded and ready for a cruise. (Courtesy of Great Lakes Historical Society.)

Bon Voyage *SS Seeandbee*! (Courtesy of Cleveland Public Library.)

Passengers supervise the tug *Kentucky* turning the *SS Seeandbee*. (Courtesy of Great Lakes Historical Society.)

The tug *Kentucky* helps to maneuver the *SS Seeandbee* into position to set sail. (Courtesy of Cleveland Public Library.)

Pollution, compliments of the *SS Seeandbee*. (Courtesy of Great Lakes Historical Society.)

"Just see that smoke so black roll from that old smokestack" of the *SS Seeandbee*. (Courtesy of Cleveland Public Library.)

Staff members are pictured here in the lobby of the *SS Seeandbee*. (Courtesy of Great Lakes Historical Society.)

This is the entertainment aboard the *SS Seeandbee*. (Courtesy of Great Lakes Historical Society.)

This was the dining room of the *SS Seeandbee*, pictured ready for diners. (Courtesy of Great Lakes Historical Society.)

Dinner is about to be served aboard the *SS Seeandbee*. (Courtesy of Great Lakes Historical Society.)

This was the lobby of the *SS Seeandbee*. (Courtesy of Great Lakes Historical Society.)

The wheelhouse of the *SS Seeandbee* is pictured here. (Courtesy of Great Lakes Historical Society.)

This photo of a cabin on the *SS Seeandbee* shows bunks, a telephone, sink, and under-the-bunk storage. (Courtesy of Great Lakes Historical Society.)

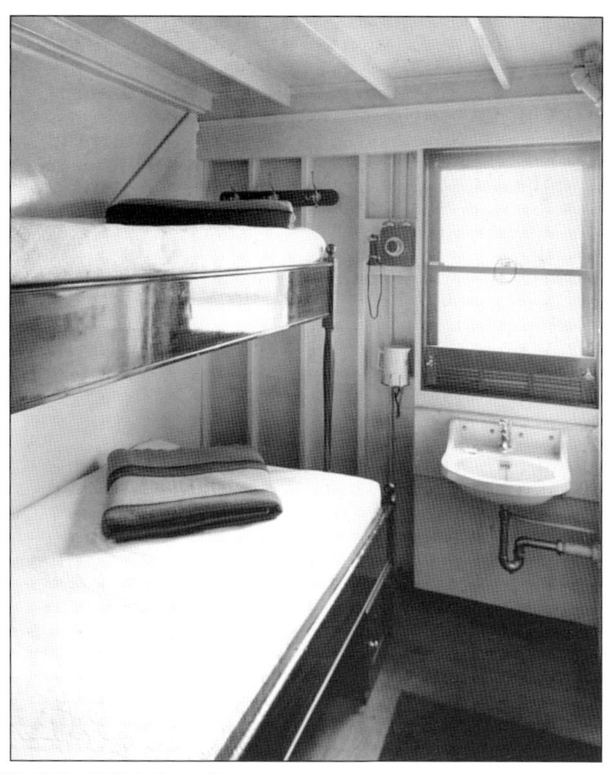

On the upper deck of the *SS Seeandbee*, the deck chairs were at the ready. (Courtesy of Great Lakes Historical Society.)

Athletes and spectators concentrate on a game of shuffleboard and a ship's officer aboard the *SS Seeandbee*. (Courtesy of Great Lakes Historical Society.)

Guests are seen enjoying the upper deck of the *SS Seeandbee*. (Cleveland Public Library.)

Guests are seen relaxing in luxury accommodations aboard the *SS Seeandbee*. (Courtesy of Great Lakes Historical Society.)

Postcard showing the atrium of the *SS Seeandbee*.

Postcard showing the main dining room of the *SS Seeandbee*.

Postcard of the SS *Seeandbee*.

Postcard of the SS *Seeandbee*.

Postcard of the SS *Seeandbee*.

Postcard of the SS *Seeandbee*.

Postcard showing the grand salon of the SS *Seeandbee*.

Three

SS GREATER BUFFALO

Owner: Detroit and Cleveland Navigation Company
Builder: American Ship Building Company, Lorain, Ohio 1923
and Great Lakes Engineering Works, Ecorse, Michigan, 1924
Dimensions: 535 feet x 58 feet 3 inches x 23 feet 7 inches

While the SS *Seeandbee* was being converted, the decision was made to have two training carriers on Lake Michigan. Had there been a sister ship to the *Seeandbee*, the choice would have been easy, but due to the provisions of the LaFollette Seamen's Act and the depressing effect on lake cruising of the 1912 capsizing of the cruise ship *SS Eastland*, plans for such a vessel had been abandoned.

Instead, the U.S. Navy turned to another ship designed by the same architect who designed the *Seeandbee*, Frank E. Kirby. This ship, the *SS Greater Buffalo*, unlike the *Seeandbee*, was owned by a financially sound company, and so it had not been considered when the plan was to have just one aircraft carrier on Lake Michigan. By 1932, the *Greater Buffalo* and its sister ship, the *SS Greater Detroit*, were the largest side-wheeled passenger boats in the world. Like the *Seeandbee*, the *Greater Buffalo* was coal-fired.

In designing the *Greater Buffalo*, Kirby had been assisted by Herbert C. Sadler, D.S.C., naval architect and professor in the naval architect and professor in the naval architecture and marine engineering department at the University of Michigan. These two had worked together on the designs of many lake vessels. The completed ship was reported to have cost $3,500,000 with the furniture, silver, and dishes costing another $500,000. It took 7,615 yards of carpet to cover the decks and passenger quarters.

The first consideration for this new ship was to make it able to get into available dry docks. This limited the draft to 16 feet and the breadth to 100 feet. The design problem was further complicated by the route of the vessel: the full length of Lake Erie, 260 miles, with shallow channels at both ends. In addition to passenger traffic, consideration had to be given to automobile and express freight transport. She and her sister ship, the *SS Greater Detroit*, were, when launched, the largest passenger ships on the Great Lakes.

The finished ship was steel hulled with a wooden superstructure. Decks on the superstructure were each about three-quarters of an acre. The length of the ship was 550 feet and the width of the hull 58 feet. Width over the side paddle-wheel housings was 100 feet.

Speed, at 22 knots, was also a consideration due to the short operating season. The total annual revenue had to be earned in as short a period as eight months each year.

Passenger accommodations on the promenade deck included 105 rooms with two berths, 32 rooms with one berth, 50 rooms with two berths and a toilet, 4 rooms with two berths, a toilet, and a shower; 12 parlors with bathrooms; and 2 parlors with toilet and shower. The gallery deck included 138 rooms with two berths; 24 rooms with one berth; 76 rooms with two berths and a toilet; 4 rooms with two berths, a toilet, and a shower; and twelve parlors with bathrooms.

The upper deck had 166 rooms with two berths for a total of 625 passenger rooms. Every room's telephone on the ship was connected to a central station on the main deck. This network was independent of the ship's inter-communicating telephones. In addition, there were rooms for housing the more than 300 officers and men.

The *Greater Buffalo* had rudders at each end of the ship to facilitate handling of the ship at each end of the run. The hull of the *Greater Buffalo* was made of steel with 11 watertight compartments in the length of ship. These compartments were formed by steel bulkheads extending from keel to main deck. The double bottom was divided into 16 watertight compartments. Where necessary, the bulkheads had watertight access doors equipped with hydraulic closing gear operated from the engine room.

The steel casings around the stacks and the vent shafts to the engine room and galley were an important part of the support of the upper decks. They formed hollow pillars from the solid steel main deck through the upper decks which were somewhat dependent on them for support. A network of steel stringers, beams, and pillars under the promenade deck also contributed to the support of the upper decks.

Accommodations for the deck, engineers', and stewards', departments were provided on the steel orlop, or lowest, deck both fore and aft of the machinery which was forward of amidships. Freight space was provided on the fore part of the main deck. The lobby entrance, cafeteria, and dining room were aft of the galley and pantry on the orlop deck.

The wooden superstructure was made of white pine. Decks and partitions were made of California redwood and the support timbers were made of Oregon fir. Bulkheads between rooms were of redwood boards mounted diagonally and glued together. Partitions in passages were of paneled white pine and exposed walls in public areas were paneled in selected hardwoods. Ceilings consisted of composition panels and ornamental plaster.

The amount of plumbing was substantial as all rooms had running water. Many rooms had baths, toilets, and showers, and some had hot and cold water. Drinking water required piping distilled water throughout the ship. Raw water was carried in 36,000 gallon steel tanks and there were 8,000 gallon tanks for sterilized water which, after being chilled, was pumped to drinking fountains throughout the ship. Water sterilized by violet ray systems was piped to all wash basins. Hot water was carried to parlors, officers' quarters, and public lavatories. There was duplicate pumping equipment for all plumbing.

Because the greater part of the operating season was in the summer, sheet metal ducts led to all inside rooms. The engine, boiler casings, the underside of the decks, and other parts of the structure that might conduct heat were insulated with magnesia board covered with galvanized sheet metal.

On the other hand, heat was provided for the early and later months of the cruising season by radiators in the public rooms and a heating pipe that ran through the outer rooms and wing passages.

Toilet spaces were ventilated with stacks; ventilator heads exhausted the foul air. Fans driven by motors on the main deck drove air through washers into the ventilating ducts which distributed it to all parts of the ship.

Fire protection was provided by a sprinkler system with heads in all staterooms, public rooms and passages, main cargo spaces, quarters below deck, and in the pilot house. Fire detectors, which automatically alarmed a watchman of any rise in temperature at any point, were installed all over the ship. The underside of the promenade deck was insulated with galvanized sheet steel, and fire doors in the cargo space and in passages leading to the rooms divided the ship into zones.

Also contributing to safety were the steel hulls with double bottoms which were divided into watertight compartments. Fifty percent more life saving features than were required by the United States government were provided including life preservers, steel lifeboats, rafts, and floats. The ship was also equipped with a wireless system with two operators always on duty.

The navigating equipment, in addition to the usual compasses, included a Sperry gyrocompass, a Sperry log, and a Haynes automatic sounding machine. The pilot house was eight decks above the surface of the water and the bridge was extended over the water on both sides to give the officers an unobstructed view along the side of the ship and beyond the stern. High-powered searchlights were located at the ends of the bridge.

In addition to life rafts and floats, 12 metal lifeboats, each with a 60-person capacity, were carried well aft, leaving the forward portion of the boat deck clear for passengers to gather to observe the ship nearing port.

Three 100-kilowatt turbo-generators, located on the main deck, provided light and power for the ship. There were about 5,000 lights on the ship and operations of the galley used 120 horsepower of motors.

--There were two 6,000-pound anchors forward and one 6,000-pound anchor aft. They were attached to 90 fathoms of two-and-one-eighths-inch cast steel-studded chain. There were seven eight by one inch mooring engines, each with 90 fathoms of one-and-one-eighths-inch diameter steel wire for making fast to the dock.

The *Greater Buffalo* was powered by three double-end and six double-end boilers fitted with forced draft and super heaters. These supplied steam for a three-cylinder compound-inclined engine-driving feathering wheels. The boilers were 14 feet in diameter; the single-end boilers were 10 feet 6 inches long and the double-end boilers were 20 feet 6 inches long. The furnaces were 54 inches in diameter and the tubes were two-and-three-fourths inches in long. All materials produced a pressure of 167 pounds in accordance with the rules of American Bureau of Shipping and the United States Steamboat Inspection Service. The boilers were in three separate compartments. There were four fire holds and four bunkers. Two ash ejectors were installed in each fire hold. The engine generated 12,000 horsepower.

The dimensions of the main engine cylinders were: low pressure 96 inches, high pressure 66 inches, stroke 108 inches. The engine framing consisted of six forged-steel struts.

Wheels were of the feathering type, designed to operate at 30 revolutions a minute. They had an outside diameter of 32 feet 9 inches with 11 floats and 14 feet 10 inches long by 5 feet wide of curved steel construction.

All this for $3,500,000!

As in the case of the *Seeandbee*, all the ornate fittings would have to be scrapped to make way for the flight deck.

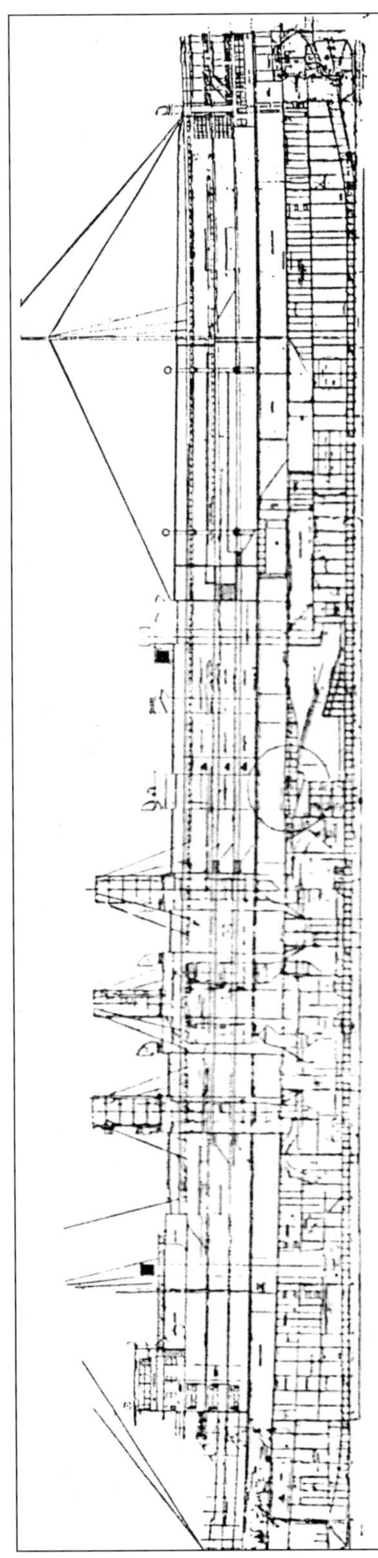

Great Lakes Side Wheel Passenger Liner Greater Detroit

Name of Vessel—GREATER DETROIT, sister ship GREATER BUFFALO.

Owner—Detroit & Cleveland Navigation Co.

Builder—American Shipbuilding Co.

Naval Architect—Frank E. Kirby.

When Launched—GREATER DETROIT Sept. 15, 1923; GREATER BUFFALO Oct. 27, 1923.

Classification—American Bureau of Shipping. *100.

HULL PARTICULARS

Length overall, 535 feet; length between perpendiculars, 519 feet; breadth molded, 58 feet hull, 96 feet 6 inches over guards; depth molded, 23 feet 7 inches, 22 feet 10 inches at guard; draft loaded, 15 feet 6 inches; displacement loaded, 9315 net tons; gross tonnage, 7739.99; net tonnage, 3330; passenger capacity, first 1194; cargo capacity, 1200 tons (short); bunker fuel capacity, 582.35 tons (at 40 cubic feet); speed, 21 miles maximum.

MACHINERY PARTICULARS

Main Engine—Name of builder, American Shipbuilding Co.; number one; type, inclined, 3-cylinder compound; size, 66 x 96 x 96 and 108-inch stroke.

Boilers—Number 3 double ended, 6 single ended; name of maker, American Shipbuilding Co.; type, Scotch; size, double ended 14 feet 2 7/16 inches mean diameter, and 20 feet 6 inches long, single ended 14 feet outside diameter and 10 feet 6 inches long; fuel, coal.

AUXILIARY EQUIPMENT

Manufacturers of:

Pumps—American Shipbuilding Co., Warren Steam Pump Co., Morris Machine Works, Dean Bros. Steam Pump Co., Union Steam Pump Co.

Windlasses—American Shipbuilding Co.

Winches—American Shipbuilding Co.

Steering Engine—American Shipbuilding Co.

Paddle Wheels—American Shipbuilding Co.

Refrigerating Machinery—Brunswick-Kroeschell Co., with Prosser steam engine from Chandler-Taylor Co., Indianapolis.

Superheaters—Power Specialty Co.

Electric Generators—Terry turbine, Allis-Chalmers generator.

Forced Draft—American Blower Co.

Gyro-Compass—Sperry Gyroscope Co.

Outboard profile and specifications of the *SS Greater Buffalo*. (Courtesy of *Marine* Review.)

Passengers are pictured checking the progress of the *SS Greater Buffalo*. (Courtesy of Great Lakes Historical Society.)

Most of the passengers are watching the nearby shore from the deck of the *SS Greater Buffalo*. (Courtesy of the Great Lakes Historical Society.)

Bow shot of the *SS Greater Buffalo*. (Courtesy of the Great Lakes Historical Society.)

SS *Greater Buffalo* is seen off on a cruise. (Courtesy of the Great Lakes Historical Society.)

SS Greater Buffalo is seen returning from another successful cruise. (Courtesy of the Great Lakes Historical Society.)

BUFFALO DIVISION

MAY 25 TO OCTOBER 1*

Fare		E.S.T.
$6.00 ONE WAY with berth (inside)	Lv. DETROIT, daily Foot of Third St.	5:30 P.M. Weekdays 5:00 P.M. Sundays and holidays
$5.00 ONE WAY without berth	Ar. BUFFALO following day	8:30 A.M. Ex. Mon. 8:00 A.M. Monday
$10.00 ROUND TRIP with berth (inside)	Lv. BUFFALO daily Foot of Main St.	6:00 P.M. Daily
$8.00 ROUND TRIP without berth unlimited stopover	Ar. DETROIT following day	9:00 A.M.

RATES FOR DELUXE BEDROOMS ON CLEVELAND AND BUFFALO DIVISIONS

Special Accommodations	Greater Detroit and Greater Buffalo	Eastern States and Western States	City of Detroit III	City of Cleveland III
OUTSIDE ROOMS:				
Single upper and double lower berth, toilet and shower......	$6.00	.....	.....	.....
DE LUXE BEDROOMS:				
Double bed and toilet..........	.....	$6.00	.....	.....
Twin beds and toilet...........	.....	6.00	$6.00	.....
Twin beds, toilet and shower....	.....	.....	.....	$6.00
Twin beds, toilet and bath......	.....	8.00	.....	7.00
Double bed, shower and toilet...	9.00	.....	.....	.....
Twin beds or double bed, toilet and bath..................	.....	10.00	.....	.....
Double bed, toilet and bath.....	.....	.....	7.00	.....

This is a price schedule for the SS *Greater Buffalo's* cruises.

Postcard of the *SS Greater Buffalo*.

Postcard of the SS *Greater Buffalo*.

Postcard of the *SS Greater Buffalo*.

This postcard of the *SS Greater Buffalo* shows her passing her sister ship, the *SS Greater Detroit*.

This postcard shows the main forward staircase of the *SS Greater Buffalo*; note the painting of Niagara Falls.

This postcard shows the dining room of the *SS Greater Buffalo*.

This postcard shows the main saloon and staircase of the *SS Greater Buffalo*.

This postcard shows the forward deck and pilot house of the *SS Greater Buffalo*.

Four

THE CARRIERS

The Navy acquired the *SS Seeandbee* in 1942 for $750,000 and estimated that the conversion to an aircraft carrier would take four months. The conversion actually took 59 days!

The conversion was begun in Cleveland with the removal of the five magnificent (mostly wood) upper decks with their contents of settees, davenports, bed springs, carpet, pillows, life jackets, basins, and bar stools, on April 14, 1942. The hull was then taken to the American Shipbuilding Company's yard at Buffalo, New York. Conversion involved constructing a massive flight deck and surrounding catwalk.

The deck was a mere 26 feet above the water, less than half the height of conventional carriers, but it was 98 feet wide to cover the paddle wheels. This was wider than the deck of the fleet carrier *USS Essex*. There was no catapult, but there was an eight-wire arrestor system to keep planes from running off the bow into the chilly waters of Lake Michigan. The employees of American Shipbuilding at Buffalo had no experience with arrestor wires, so they rigged a system which they thought would be adequate and ran into it with loaded steel-hauling trucks. The system worked successfully through the life of the ship. No lifts, catapults, or hangar facilities were required on the ship; the planes were over-nighted at Naval Air Station, Glenview. The main deck area included officers' quarters aft, wardroom, and crew's quarters forward, as well as a motion picture equipped instruction room, a ship's service store, laundry, tailor shop, barber shop, dining room, and crew's recreation room. A small "island" was constructed over the starboard paddle wheel sponson to provide realistic flight deck appearance. This structure housed the four funnels and limited command facilities. It also gave a "feel" to the flight deck similar to that of a fleet carrier. The converted ship carried no armament, no armor, and no radar.

When the conversion was completed, the *Wolverine* had the following specifications:
Displacement: 6,381 tons
Length of Hull: 500 feet
Beam: 99 feet
Draft: 15 feet 5 inches
Length of Flight Deck: 558 feet 6 inches
Six single-end and three double-end coal-fired boilers
Side-wheel paddles, each 32 feet in diameter, with 11 feathering blades on each wheel
Maximum speed: 17 1/2 knots
Average fuel consumption: 150 tons of bunker coal in 24 hours at full speed
Ship's Complement: 22 officers, 300 enlisted men

No modifications of the *Seeandbee's* mechanical system were found necessary.

The conversion was completed on August 11, 1942, and the ship was commissioned under the command of Commander G.R. Fairlamb and christened the *USS Wolverine* on August 12, 1942.

The name *Wolverine* was chosen to honor the State of Michigan where the ship had been constructed, to recognize that the ship would be sailing on Lake Michigan, and to commemorate the world's first iron-hulled ship commissioned in 1844 and decommissioned in 1942.

On August 22, 1942, Chicago Mayor Edward J. Kelly and a large group of government and naval dignitaries welcomed the *Wolverine* to the Chicago area. Lake Michigan was somewhat choppy that day, so the welcoming ceremonies were held on the *USS Wilmette*.

USS Wilmette

The Naval Reserve training ship *USS Wilmette*, a local fixture of the Chicago lakeshore through the 1920s, 30s, and 40s, was a refloated and refitted *SS Eastland*. The Great Lakes cruise ship had capsized in the Chicago River on July 24, 1915, when the Western Electric Company had reserved several lake cruise ships to ferry employees and their families from Chicago to Michigan City, Indiana for a company picnic.

Because the *Eastland* was the largest and best known of the chartered ships, there was a rush of picnickers to board it, seriously overloading the ship. When a Chicago fireboat came down the river, many of the passengers ran to the river side of the still-moored *Eastland* to get a better look. The shift in weight caused the ship to capsize.

The tragedy took the lives of 841 passengers, two crewmen, and a member of the crew of the nearby *SS Petoskey*. By contrast, the better-known sinking of the *RMS Titanic* three years before had taken the lives of 694 crew members, but only 829 passengers.

It is ironic that the capsizing of the *Eastland* was the event that convinced the builders of the *Seeandbee* not to build a sister ship.

The intertwining of the activities of the *Wilmette* and the *Wolverine* continued when the *Wilmette* joined the group of ships which would constitute the Chicago fleet.

After the welcome, the shakedown had problems. A company of engineers riding the ship wrote, "The crew still needs considerable experience in throwing the coal the full length of the furnace." It must be remembered that the Navy had no other coal-fired ships, and so it had no experienced stokers. The crew also had to get used to the idea of taking boiler-feed water directly from Lake Michigan, which was possible because the lake was fresh water.

The skipper originally had doubts about the boilers and had recommended a conversion to oil. He withdrew this request after some operating experience, noting that "using high grade Pocahontas coal and with increased boiler room efficiency, the *Wolverine* now operates most satisfactorily from the standpoint of smoke nuisance."

Before the commissioned and christened *Wolverine* had left the shipyard at Buffalo, the *SS Greater Buffalo* had arrived, and its conversion had begun. Techniques and procedures learned from the conversion of the *Wolverine* were employed in the conversion of the *Greater Buffalo*.

The most dramatic difference between the two ships was that the *Wolverine* was fitted with an oak flight deck in common with the fleet carriers at the time. The *Greater Buffalo*, which was renamed the *USS Sable*, had a flight deck made of two designs of steel flight decking topped with eight types of commercial non-skid coatings, applied in a checkerboard pattern.

Sable's conversion was not completed before winter made navigation on the Great Lakes impossible, so it did not join the *Wolverine* until spring of 1943. She was commissioned at Buffalo on May 8, 1943.

When converted, the Sable's specifications were:
Displacement: 8,000 tons
Length of Flight Deck: 535 feet
Draft: 15 feet 5 inches
Machinery: 4 coal-fired boilers
Horsepower: 10,500
Speed: 18 knots

Here, the *SS Seeandbee* is seen being converted to the *USS Wolverine*. (Courtesy of U.S. Naval Historical Center.)

SS Seeandbee is seen here on her way to becoming the *USS Wolverine*. (Courtesy of U.S. Naval Historical Center.)

USS Wolverine is seen here running official trials off Buffalo, New York on August 11, 1942. (Courtesy of U.S. Naval Historical Center.)

The just-completed *USS Wolverine* is pictured here. At left is the stern of the *SS Greater Buffalo* readying for conversion to the *USS Sable*. (Courtesy of U.S. Naval Historical Center.)

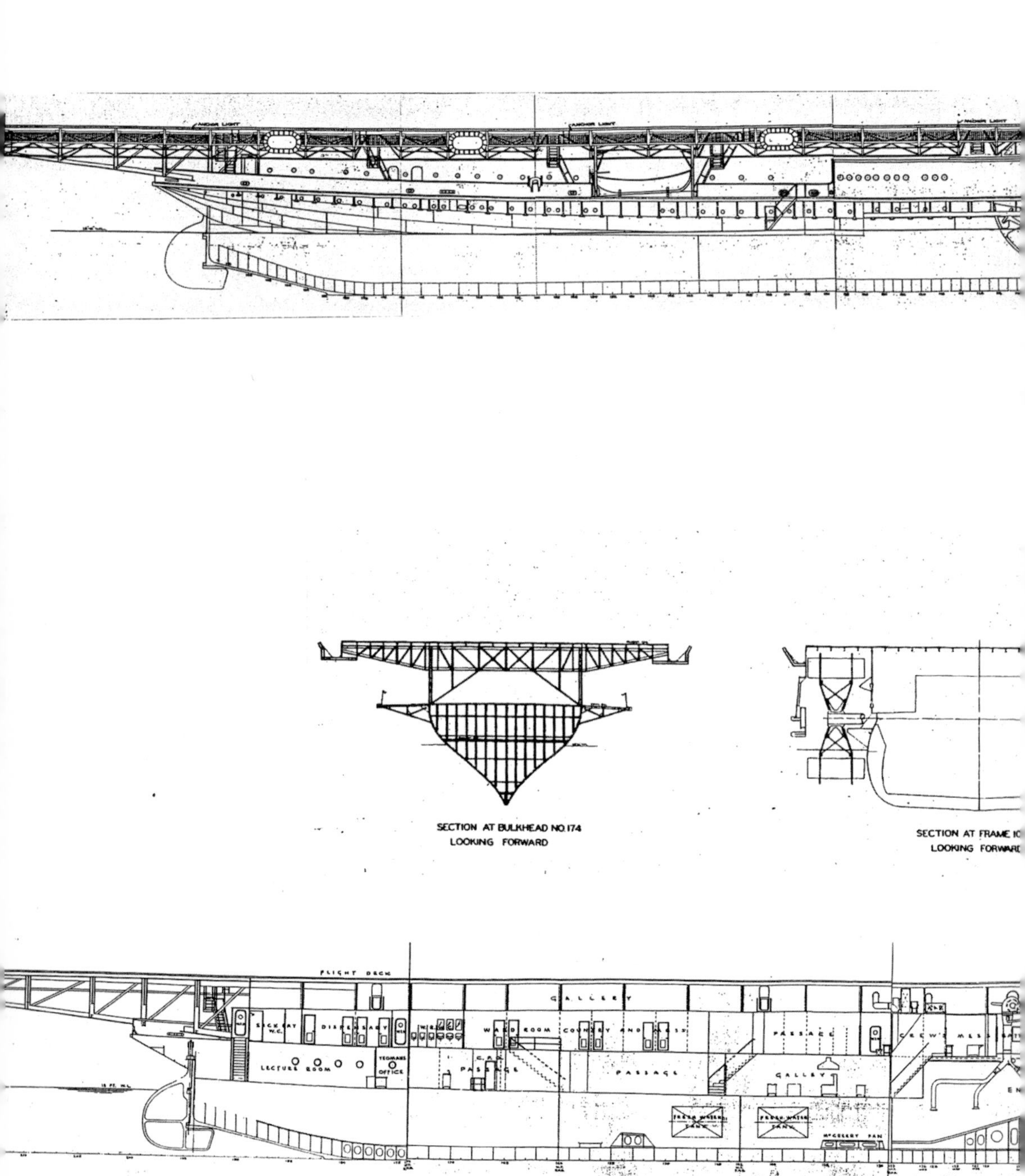

These are the outboard and inboard profiles and cross sections of the *USS Wolverine*.

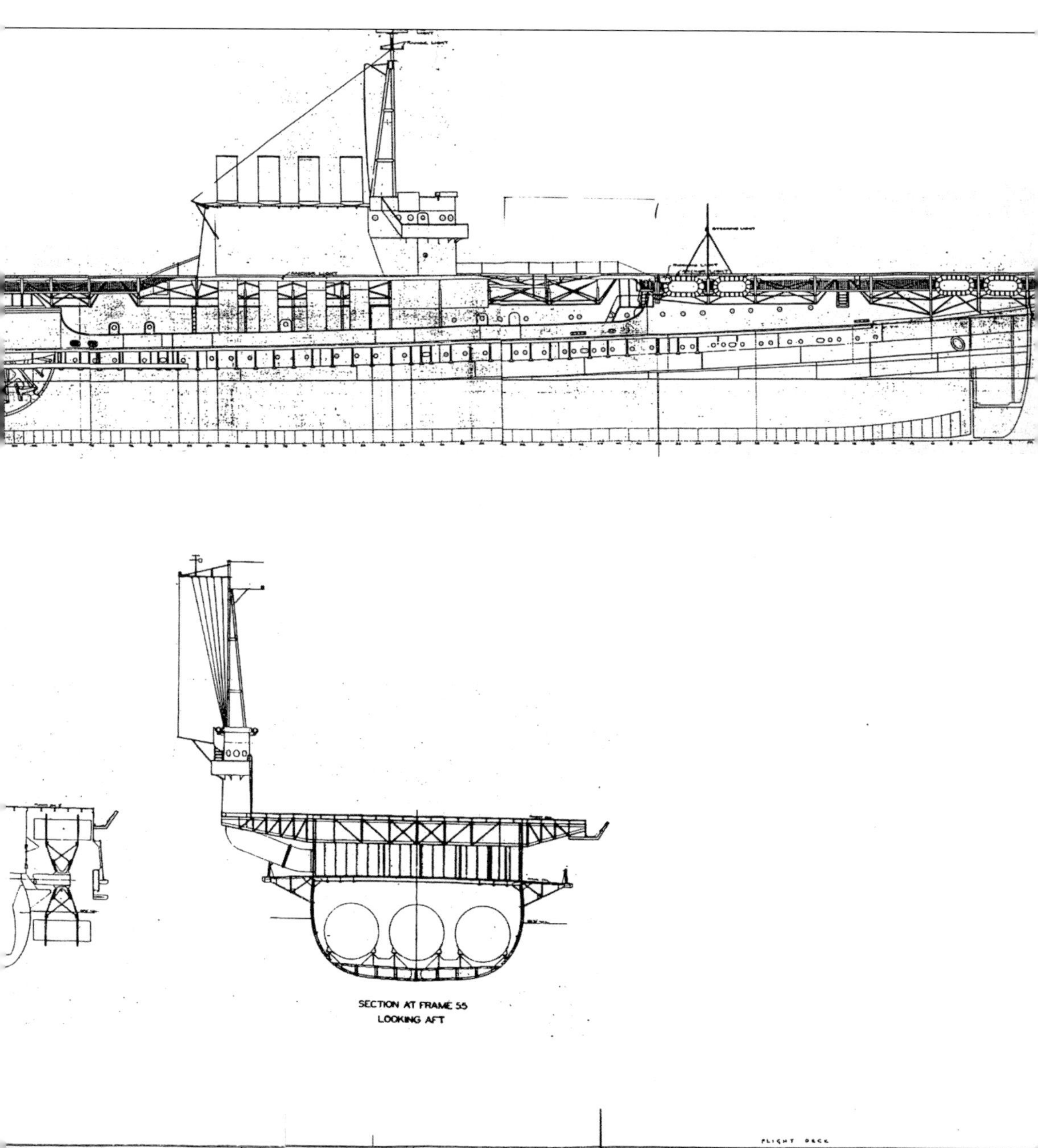

SECTION AT FRAME 55
LOOKING AFT

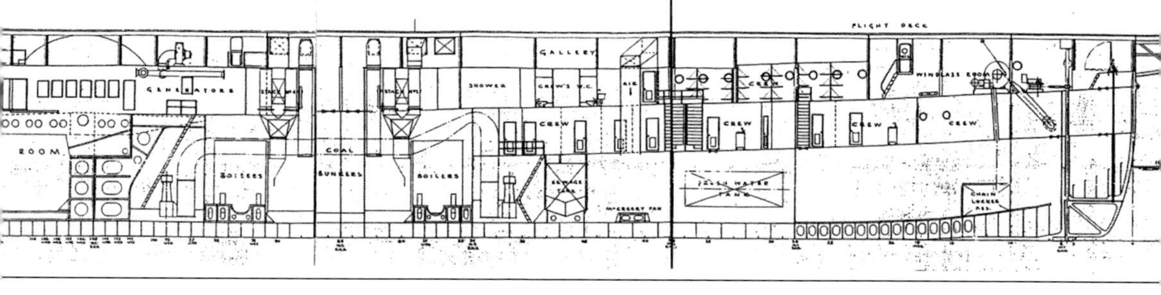

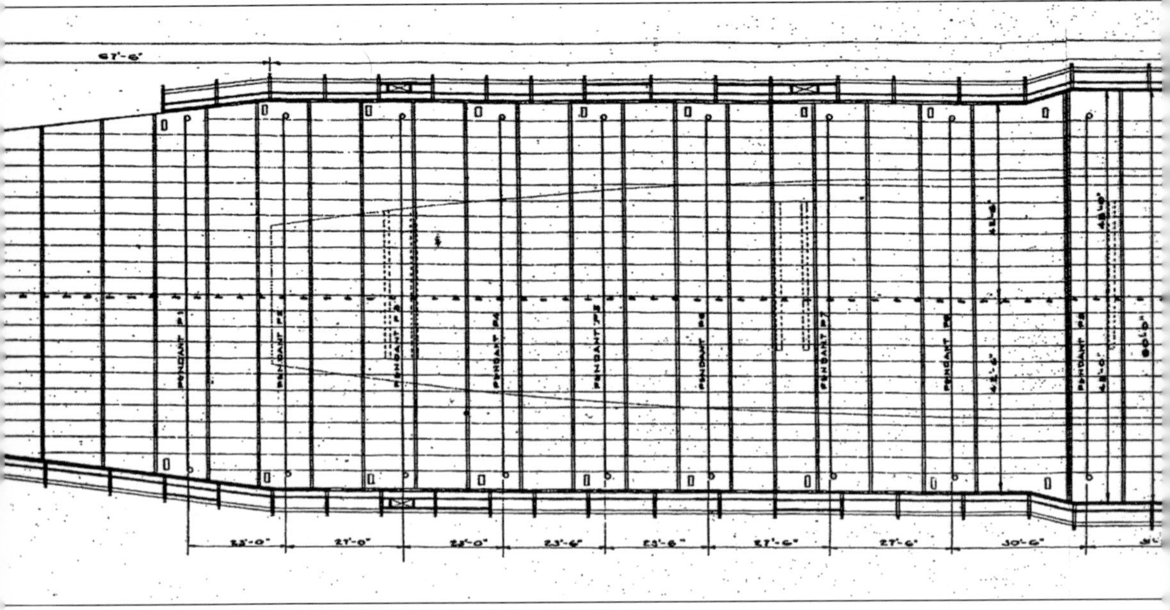

This is the outline of the flight deck of the *USS Wolverine*.

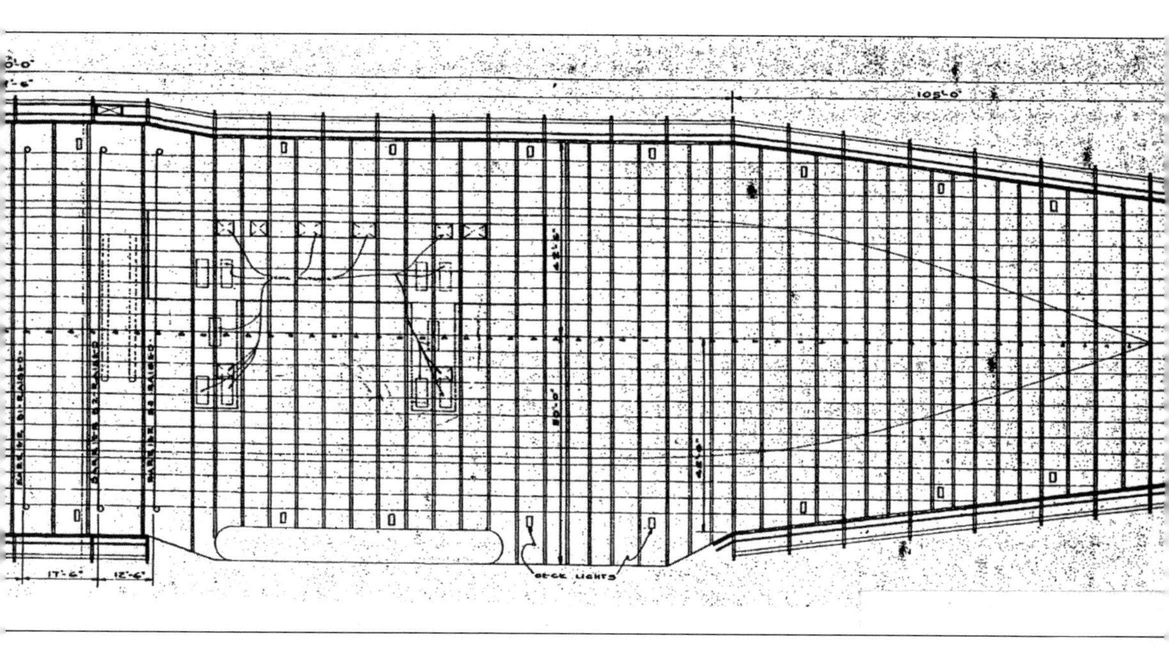
105'-0"
17'-6"
12'-6"
DECK LIGHTS
80'-0"

This photo shows the *USS Wolverine*. (Courtesy of the Buffalo Police Department.)

USS Wolverine is seen here on the date of her commissioning at Chicago, Illinois, August 22, 1942. (Courtesy of U.S. Naval Historical Center.)

SS Eastland is seen here in its original service as a Great Lakes cruise ship. (Courtesy of Great Lakes Historical Society.)

SS Eastland capsized in the Chicago River on July 24, 1915. (Courtesy of Great Lakes Historical Society.)

USS Wilmette is pictured here after its conversion from the *SS Eastland*. (Courtesy of the National Archives.)

This image shows a pilot's eye view of the *USS Wolverine*. (Courtesy of U.S. Naval Historical Center.)

The *USS Wolverine* is seen here. (Courtesy of U.S. Naval Historical Center.)

SS Greater Buffalo en route to being converted into the *USS Sable*. (Courtesy of U.S. Naval Historical Center.)

This photo shows the fitting out of the *USS Sable* at Buffalo, New York, during conversion from the *SS Greater Buffalo*. (Courtesy of U.S. Naval Historical Center.)

This is a stern view of the *USS Sable*. (Courtesy of Great Lakes Historical Society.)

This is a head on view of the *USS Sable*. (Courtesy of Great Lakes Historical Society.)

The *USS Sable* is pictured underway on the Great Lakes during June 1945. (Courtesy of the U.S. Naval Historical Society.)

Five

OPERATION

On August 1, 1942, the Carrier Qualification Training Unit (CQTU) was commissioned at Naval Air Station, Glenview, under the command of Cmdr. E.J. O'Neill. Its mission was to equip pilots with the skills needed for aircraft carrier landings and take-offs using the newly-commissioned *USS Wolverine*. Because there were no aircraft storage facilities aboard the Wolverine, the next order of business in training carrier qualified pilots was to find a land-based home for the planes and pilots.

The landing field at Naval Air Station, Glenview, provided the first site for take-offs, landings, overnight storage, repairs, and student and staff housing. Because Glenview had other programs using its runways, the field at the nearby Douglas Aircraft plant was also used. This site had drawbacks as Douglas used the field for testing the DC-4 (aka C-54) transports which the plant was manufacturing. Incidentally, this facility which became the site for O'Hare International Airport, was, at the time, called Orchard Place, which is why O'Hare-bound luggage is tagged ORD.

In addition to customary take-offs and landings, the field used for carrier training needed to have an area dedicated to simulated carrier operations. Thus, it needed a field of its own. A suitable field was located about nine miles northwest of Glenview. This field, known as Allendale (later Libertyville), was used from its preparation through December 1942. Because the field had become pot-holed and rutted, improvements of the Allendale field or the acquisition of a new site were deemed necessary. A new site near Elgin, Illinois was located and acquired.

Operations continued until the winter of 1942, when severe weather conditions caused a curtailment of all activity. The *Wolverine* was icebound five miles offshore for two days. By December 22, 1942, it was moored safely alongside Navy Pier, having qualified 287 pilots since the beginning of the program.

On December 24, 1942, the unit received orders to transfer to Naval Air Station, San Diego, where qualifications were to continue using the *USS Long Island* and the *USS Core* (like the *Wolverine*, these were converted civilian vessels). Though costly, this move took only 11 days. During the California experiment, 240 additional pilots were qualified. A comparison of the costs with the results caused the Navy to decide that operations at Glenview would be continued year-round on Lake Michigan.

Contributing to the decision not to return to California during subsequent winters were the facts that the carriers operating out of San Diego were forced to operate 100 or more miles out to sea, with armed escorts, and under wartime restrictions which were unnecessary on Lake Michigan; the ships returned to port on weekends for required maintenance and repairs. Also, the *Long Island* and the *Core* were short and tended to pitch, which added to the difficulties for pilots just learning carrier landings and take-offs. Further, the wind was no easier to find on the Pacific Ocean than it was on Lake Michigan, and living accommodations were harder to find in San Diego.

Operating off the Pacific coast required radio silence, and armed and armored escort vessels which were unnecessary when the unit was at Glenview. Flyovers by PBY's were not uncommon. The most noticeable advantage at San Diego was the ready availability of replacement parts for the regular Navy ships; an advantage not enjoyed by the coal-fired, side-wheeled converted lake cruisers in Chicago.

While the unit was based at San Diego, little more than thinking about alternative fields could be done. Because of northern Illinois winters, being what they are, earth-moving and related construction were not feasible. Thus, when the unit returned to Glenview in March of 1943, nothing had been changed at Allendale but the decision had been made not to use the Elgin site. Construction at Allendale included runways and arresting gear was completed on September 23, 1944.

The Wolverine resumed operation on St. Patrick's Day—a full two months before the opening of the Great Lakes shipping season. This ended the California experiment. During May of 1943, the *USS Sable* joined the Chicago fleet which meant that pilot qualification could continue uninterrupted even when one of the ships was off the lake for repairs, overhauling, or coaling. Cabin cruisers, *Lark* and *Peregrine*, and a freight lighter, *Commerce*, were the regular escort vessels required on Lake Michigan; none of them needed to be armed or armored. *Lark* and *Peregrine* were available to rescue students whose plane missed their mark and made a fresh-water landing in Lake Michigan. *Commerce* served as a tender, ferrying personnel, supplies, and crippled planes between the carriers and shore.

From time to time, Coast Guard cutters, a Coast Guard icebreaker, and the *USS Wilmette* completed the fleet. The ice breaker was necessary because the hulls of both the *Wolverine* and the *Sable* had been constructed to engage in ice-free summer cruises.

The carrier pilots' training program at Glenview included one day of classroom work with a Landing Signal Officer, one day of field landing practice on land (first, in an advanced trainer—usually the Navy's SNJ—then, in the pilot's assigned combat aircraft), and then one day of carrier landings.

On August 25, 1942, Lt. Cmdr. Eugene J. O'Neill, skipper of the CQTU, became the first man in history to land on a fresh-water, coal-fired, paddle-wheeled aircraft carrier. On September 12, 1942, Ensign Biedleman, USNR, became the first pilot to qualify under the program.

For carrier landing, a squadron of planes would take off and rendezvous at what the Navy called "Point Oboe" (which Chicagoans recognized as the site of the Baha'i House of Worship in Wilmette). The Squadron Leader would then make radio contact with the *USS Wolverine* or the *USS Sable* and proceed with landings. This routine operated seven days a week with an occasional day off for coaling in Calumet, near Gary, Indiana.

It continued even during the winter of 1944–45, the most severe in the history of the Chicago Weather Bureau, when Lake Michigan was frozen up to 15 miles from shore.

The carriers would get underway as early as 3:00 a.m. to avoid enveloping downtown Chicago in coal smoke. The Landing Signal Officer (LSO) would fly to his ship a half hour before the first trainees arrived. Landing started at 8:00 a.m. and continued for an hour, which would permit the aircraft to return to land before dark.

Early in the program, billowing smoke from the funnels reduced visibility, making landings difficult, or, frequently, impossible. As stokers became more proficient at pitching coal far enough into the boilers, this became less of a problem. Also, a higher grade of Pocahontas coal was used, which burned much more cleanly.

Because Lake Michigan runs basically north and south and Chicago is situated in the southwest corner of the lake, the carriers would steam north into the wind to facilitate take-offs. The *Sable* was slightly faster than the *Wolverine* and had a slightly longer flight deck with two more arrestor wires, so it was able to handle heavier aircraft. After dark, the carriers would steam south to Chicago in time for another day of qualifications.

The training squadrons received three weather reports each day. These were especially valuable during the winter months when there was a lot of fog on Lake Michigan, and the pilots did not have the carriers in sight. Radio messages directed the squadrons to the ships. Originally, students had been required to make eight landings and take-offs to qualify as carrier–qualified pilots. By the end of the program, that number had increased to 14. Qualified fighter, torpedo-bomber, and dive-bomber pilots graduated from the CQTU as qualified carrier pilots. Planes were often refueled on the flight decks of the carriers.

In the beginning, Commander Whitehead had projected the qualification of 30 pilots per day; by 1944, that rate had more than doubled. Vital statistics of the program include 136,428 landings qualifying 17,820 pilots. Twenty-one pilots were killed during training operations—none attributed to shipboard operations. The *Chicago Tribune*, in its December 31, 1969 issue, estimated that there were over 100 aircrafts at the bottom of Lake Michigan as a result of the program. On October 28, 1990, the *Chicago Sun-Times* estimated there might be as many as 300.

On August 24, 1943, an 18-year-old pilot named George H.W. Bush received his carrier qualification. He went on to a distinguished career in naval aviation and ultimately became Commander in Chief of all United States armed forces.

In addition to qualifying pilots, the *Wolverine* and *Sable* trained many Landing Signal Officers (LSO's), Flight Deck Officers (FDO's), and other deck crew members. Three additional LSO's were added to the CQTU staff when the *USS Lexington*, the *USS Yorktown*, and the *USS Hornet* were sunk by Japanese aircraft or submarines.

On August 10, 1943, the *Sable* sailed to a spot off Traverse City, Michigan, where it participated in trails of the Navy's then-secret TDN drone aircraft.

A little known distinction was the *Wolverine* becoming the flagship of the United States Navy for a day when Adm. Ernest J. King inspected the ship and broke his flag on board.

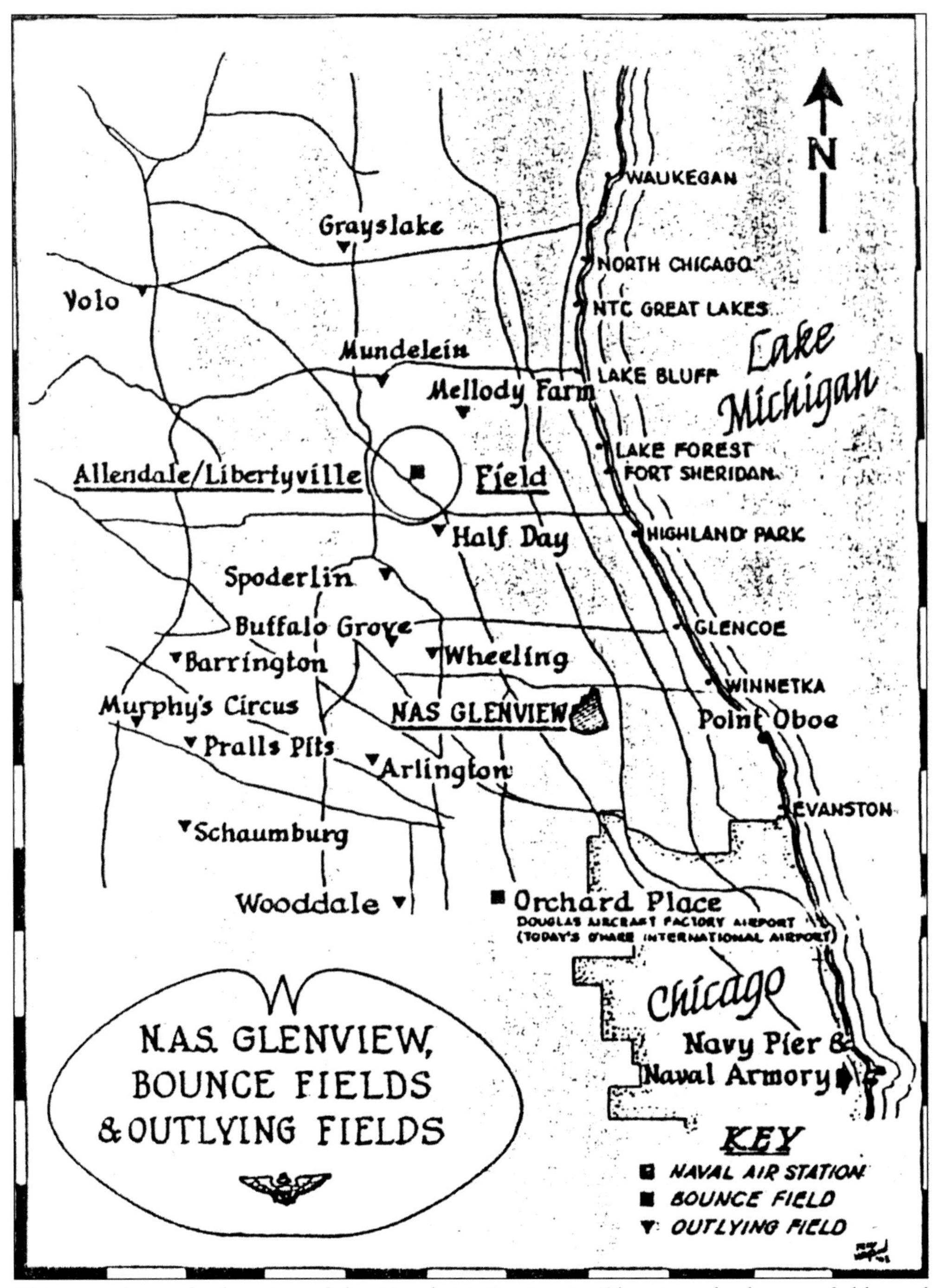

This map shows the location of the Naval Air Station in Glenview, the bounce fields, and outlying fields.

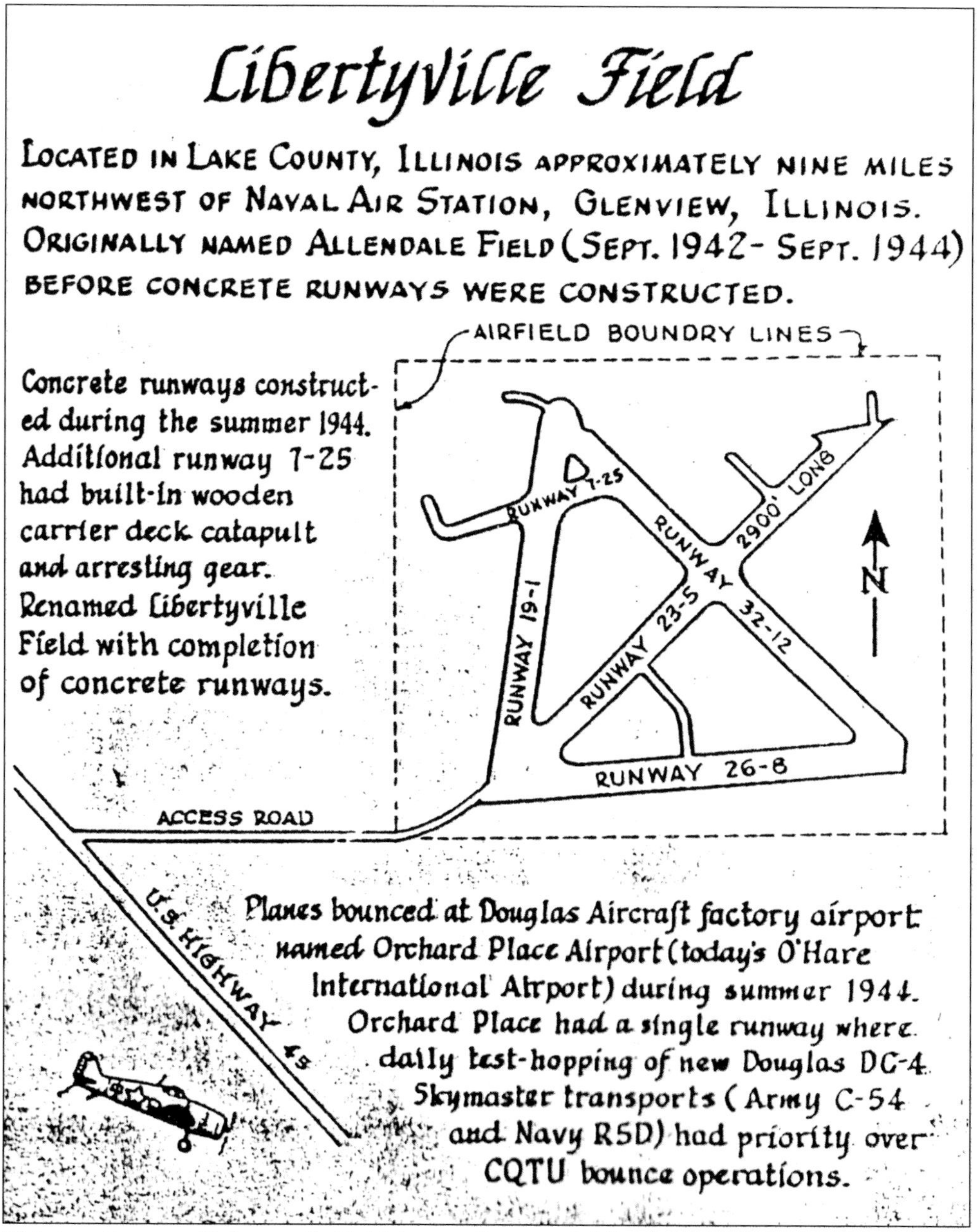

Libertyville Field

LOCATED IN LAKE COUNTY, ILLINOIS APPROXIMATELY NINE MILES NORTHWEST OF NAVAL AIR STATION, GLENVIEW, ILLINOIS. ORIGINALLY NAMED ALLENDALE FIELD (SEPT. 1942- SEPT. 1944) BEFORE CONCRETE RUNWAYS WERE CONSTRUCTED.

AIRFIELD BOUNDRY LINES

Concrete runways constructed during the summer 1944. Additional runway 7-25 had built-in wooden carrier deck catapult and arresting gear. Renamed Libertyville Field with completion of concrete runways.

RUNWAY 7-25
RUNWAY 19-1
RUNWAY 32-12
2900' LONG
RUNWAY 23-5
RUNWAY 26-8
N
ACCESS ROAD
U.S. HIGHWAY 45

Planes bounced at Douglas Aircraft factory airport named Orchard Place Airport (today's O'Hare International Airport) during summer 1944. Orchard Place had a single runway where daily test-hopping of new Douglas DC-4 Skymaster transports (Army C-54 and Navy R5D) had priority over CQTU bounce operations.

This is a diagram of Libertyville Field (originally named Allendale Field).

This is a pilot's eye view of Point Oboe, Wilmette, Illinois, which was a rendezvous point for groups of trainees heading for carriers on Lake Michigan. The white building in the center of the photo is the Baha'i House of Worship, and the broad expanse at the top is the surface of Lake Michigan. (Courtesy of Baha'i House of Worship.)

In reply address not the signer of this letter, but the Commanding Officer, USNR Aviation Base, Glenview, Ill.

NAVY DEPARTMENT
U. S. NAVAL RESERVE AVIATION BASE
CHICAGO, ILLINOIS

MAIL ADDRESS: GLENVIEW, ILLINOIS

10 DEC 1942

Mr. Harry C. Kinne, President,
Village of Wilmette,
Wilmette, Illinois.

Dear Sir:

We are glad to comply with your request that our planes refrain from flying low over the Baha'i Temple and have issued instructions to our pilots to keep clear. Any violations should be reported and appropriate action will be taken.

Sincerely,

G.A.T. Washburn

G.A.T. WASHBURN
Commander, USN,
Commanding.

This is a copy of a letter from the commander of the U.S. Naval Reserve Aviation base to the president of the Village of Wilmette regarding low-flying aircraft over the Baha'i House of Worship (Point Oboe). (Courtesy of National Baha'i Archives.)

Here, applicants for Navy carrier training are given a view of the bow of the *USS Wolverine*. (Courtesy of Milwaukee Public Library.)

USS Wolverine conducted flight training activities on Lake Michigan during 1943. The aircraft in the foreground are SNJ-3 Texan trainers. The aircraft in the background is an SBD dive bomber. (Courtesy of U.S. Naval Historical Society.)

USS Long Island, converted from *Mormacmail*, which, with the *USS Core*, handled the CQTU program out of San Diego, California, during the winter of 1942–43. (Courtesy of Naval Historical Society.)

Pictured here is the *USS Core*.

This was the first landing of an F4U on the *USS Wolverine*, April 2, 1943. This was the 500th landing on the carrier. (Courtesy of the National Archives.)

Not all pilots qualified on Lake Michigan were members of the U.S. Navy or Marine Corps. Note the insignia on this Martlet I, the British Fleet Air Arm's version of the Grumman Wildcat. (Courtesy of the Chicago Maritime Society.)

This view of the downtown Chicago skyline was taken from the deck of the *USS Sable*. (Courtesy of Frederick Clark Durant III.)

A North American SNJ-3 trainer is seen here taking off from the *USS Sable* during May 1945. Note the billowing smoke from the *Sable's* coal-fired boilers. (Courtesy of U.S. Naval Historical Center.)

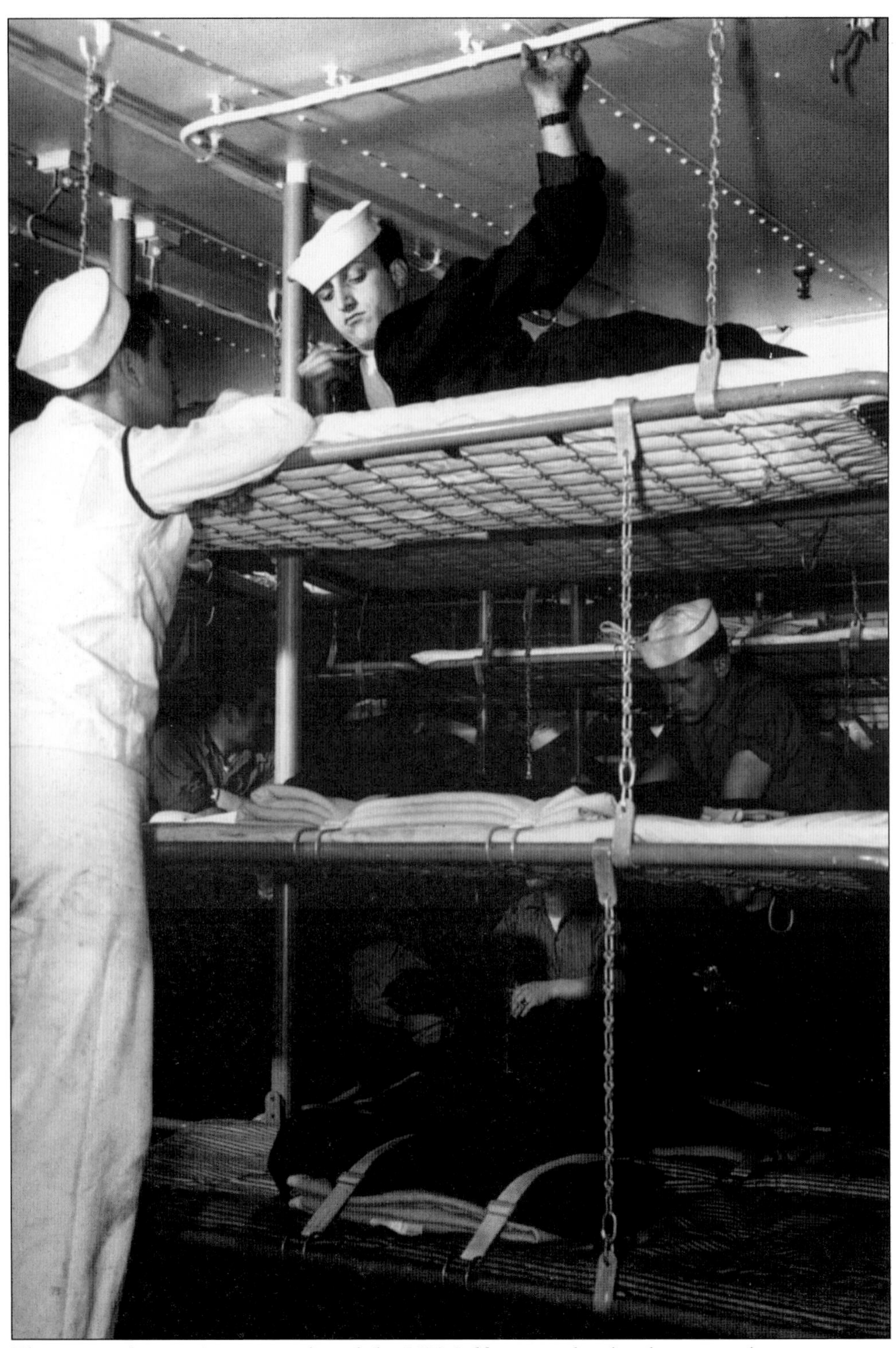

These were the crew's quarters aboard the *USS Sable*, somewhat less luxurious than staterooms aboard the *SS Greater Buffalo*. (Courtesy of the National Archives.)

Pictured is the ship's galley aboard the *SS Sable*. (Courtesy of the National Archives.)

Stoking the coal-fired boilers aboard the *USS Sable* was an exhausting task. (Courtesy of U.S. Naval Historical Center.)

A Grumman Avenger torpedo bomber is seen here aboard the *USS Sable*. Note the arrestor wire. (Courtesy of Frederick Clark Durant III.)

This photo shows the flight deck of the *USS Sable* with one North American SNJ-4 trainer and

13 General Motors FM-2 fighters. (Courtesy of the National Archives.)

This is a pilot's-eye view of the *USS Sable* on Lake Michigan. (Courtesy of Frederick Clark Durant III.)

This aircraft is going in for a landing on the *USS Sable* with arresting hook at the ready. (Courtesy of Frederick Clark Durant III.)

Pilots had little opportunity to enjoy the downtown Chicago skyline while landing and taking off from the *USS Sable*. (Courtesy of Frederick Clark Durant III.)

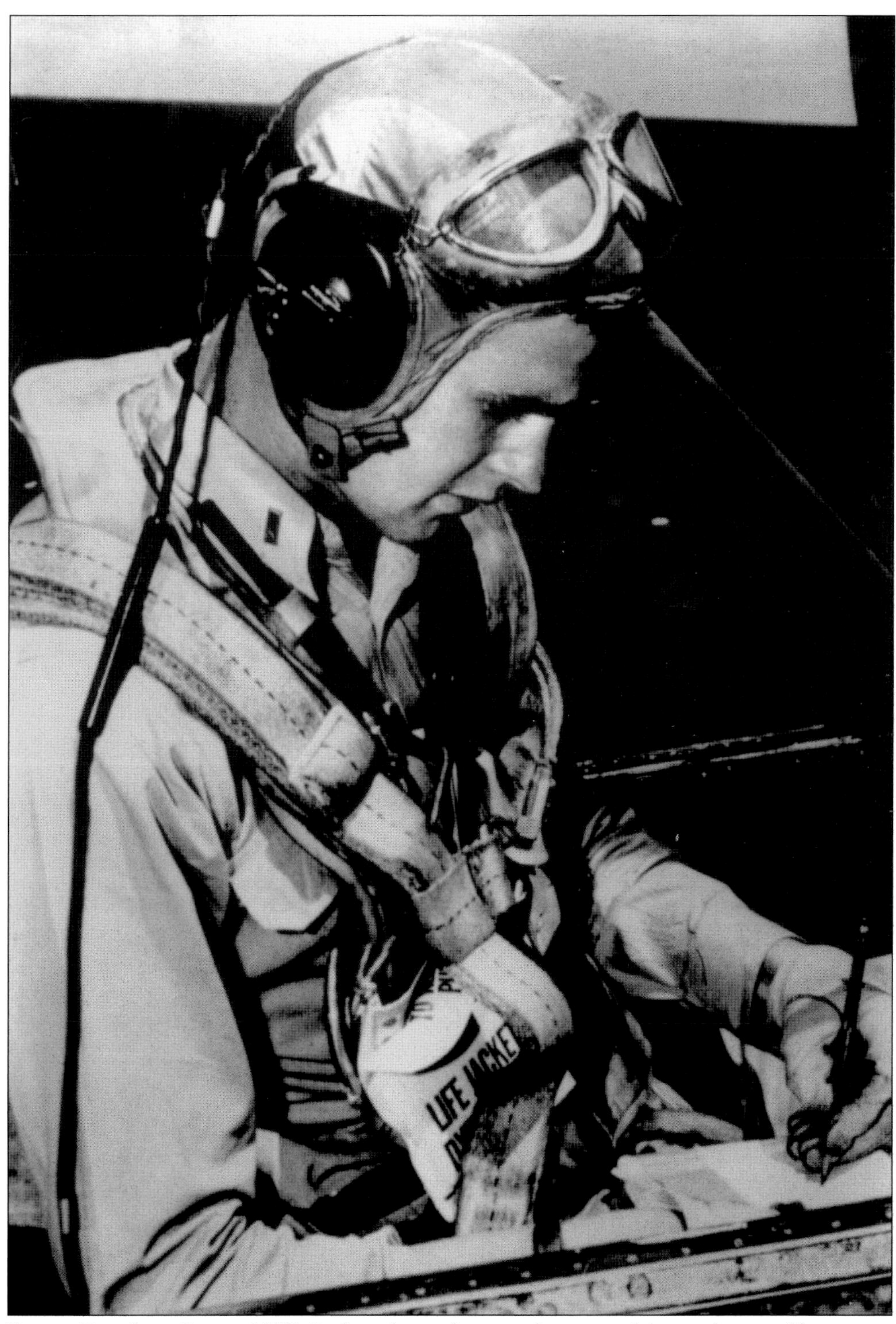

Former President George H.W. Bush is shown logging the successful completion of his carrier pilot qualification aboard the *USS Sable* on August 24, 1943. He was the second youngest pilot to qualify under the CQTU. (Courtesy of Chicago Maritime Society.)

The *Commerce*, a ship of the CQTU fleet, is seen coming alongside the *USS Sable* to remove

an FM-2 from the catwalk. (Courtesy of Frederick Clark Durant III.)

USS Sable was photographed in June 1945 after a *Wildcat* was stopped by the barrier. (Courtesy of U.S. Naval Historical Center.)

USS Sable was photographed off Traverse City, Michigan, on August 10, 1943, during tests of then-secret TDN drone aircraft. (Courtesy of U.S. Naval Historical Center.)

This crashed FM-2 fighter is pictured on the flight deck of the *USS Sable*, May 1943.

A TBN drone aircraft was being loaded aboard the *USS Sable* for tests off Traverse City, Michigan, on August 10, 1943. (Courtesy of the National Archives.)

USS Sable is seen here off Traverse City, Michigan, with a pair of TDN drones aboard for flight tests. (Courtesy of National Archives.)

Six

THE END

When Gen. Douglas MacArthur, standing on the deck of the battleship *USS Missouri* witnessing the Japanese surrender at the end on the World War II, announced, "These proceedings are at an end," he was proclaiming more than the end of the shooting and bombing war; many other realities also came to an end. Included among these was the need for large numbers of carrier-qualified navy fighter and bomber pilots. This meant that the mission of the *USS Wolverine* and the *USS Sable* had been completed.

In a letter from the Commander of Naval Air Operational Training, he wrote:

> *The* USS Wolverine *and the* USS Sable *have at all times been maintained in an efficient state of readiness and have qualified a total of 17,820 pilots for carrier duty. The skill and resourcefulness, the spirit of enthusiasm and teamwork demonstrated by all officers and men have made possible this high rate of carrier qualification commensurate with safety of operations. For the vital role in the mission of this command, the Chief of Naval Air Operations Training desires to extend to all officers and men, his appreciation for a job well done and a heartfelt and warm farewell to our Great Lakes Carriers.*

Demobilization of the ships came quickly; they were decommissioned on November 7, 1945, turned over to the War Shipping Administration and stored at "Finger Pier" at the foot of Grand Avenue near Chicago's Navy Pier awaiting disposition.

As required, the ships were offered to their original owners. Coast Guard requirements made re-conversion to cruise ships non-economical, but there was no shortage of suggestions for future use. Railroads and automobile manufacturers considered their use as ferries. As at the start of the war, their widths precluded their use of the Welland Canal on the way to ocean activity. There was even a feasibility study centered on turning them into floating night clubs—think of the immense dance floors . . . and the liability insurance, recognizing the possibility of intoxicated dancers falling into Lake Michigan! If one of the ships had survived until the late 20th century, it might well have been converted into a gambling boat.

One of the most creative suggestions came from the Lake Erie International Vacation Land Conference. They petitioned Congress to transfer the *Sable* to the Bureau of National Parks for use in Put-In-Bay as a breakwater protecting the Commodore Perry monument. However, under the terms of the Surplus Power Act, the Maritime Commission could not give the ship away, and the National Park Service did not have available funds to make even a minimum bid.

It seems a shame when navy veterans such as the *USS Constitution*, the *USS Texas*, and the *USS Alabama* were preserved thousands of miles from the areas of their contributions to America's defense that the neither the *USS Wolverine*, nor the *Sable* could be preserved at the site of their unique wartime contributions off the lake shore at Chicago.

Indeed, the most striking remembrance of these carriers is in a side room at the National Museum of Naval Aviation in Pensacola, Florida, displaying an F4F Wildcat and an SBD Dauntless recovered from the bottom of Lake Michigan. This display is breathtakingly presented but seems a modest recognition of the ships' service.

On November 28, 1947, the *Wolverine*, having been sold for $46,789 to the A.F. Wagner Iron Works, was towed from Chicago to Jones Island at Milwaukee, Wisconsin, by the tug *John Roen III*, to be scrapped.

The *Sable* continued to accrue $4,800 per month mooring expense to custodian, Hoskins Coal & Dock Co., until July 7, 1948, when it was sold to the Steel Company of Canada for $126,176. The tugs *Lachine* and *Guardian* towed the ship to Port Colborne, Ontario.

Just as the *Greater Buffalo* had been too wide for the Welland Canal, so was the *Sable*. The problem was solved by cutting the control tower and the paddle wheels from each side of the ship at Port Colborne, and placing them in the middle of the flight deck, thus narrowing the ship by 14 feet on each side. The tugs *Helena* and *James Stewart* were then able to tow the *Sable* to Hamilton, Ontario, to finish the scrapping.

Richard F. Whitehead, the man responsible for suggesting this chapter in naval aircraft history, died in 1993 at the age of 99; he was a retired Vice Admiral.

This is a photo of the *USS Wolverine* taken from the *USS Sable* after the decommissioning of both ships. (Courtesy of Frederick Clark Durant III.)

The decommissioned *USS Wolverine* is seen tied up near Navy Pier, Chicago, Illinois. (Courtesy of Great Lakes Historical Society.)

With its mission accomplished, the *USS Wolverine* bids hail and farewell to the downtown Chicago skyline as it is towed by the tug *John Roen III*, on its way to Milwaukee to be scrapped. (Courtesy of Cleveland Public Library.)

After decommissioning, the *USS Wolverine* (left) and the *USS Sable* were tied up at Finger Pier, near the foot of Grand Avenue, awaiting final disposition. (Courtesy of Chicago

Maritime Society.)

USS Wolverine is seen here at Jones Island, Milwaukee, Wisconsin, awaiting scrapping. (Courtesy of Milwaukee Public Library.)

USS Wolverine is pictured being scrapped at Jones Island, Milwaukee, Wisconsin, on April 3, 1948. (Courtesy of Milwaukee Public Library.)

The width of the Welland Canal dictated that the *USS Sable* be narrowed 28 feet on her way to being scrapped in Hamilton, Ontario. (Courtesy of Great Lakes Historical Society.)

USS Sable is seen here at Port Colborne, Ontario, being narrowed 14 feet on each side by the removal of the paddle wheel and bridge (piled on flight deck) in preparation for transit through the Welland Canal on the way to Hamilton, Ontario, to be scrapped. (Courtesy of Milwaukee Public Library.)

The wreck of the Grumman F4F Wildcat was recovered from the bottom of Lake Michigan and is now displayed in a dramatic exhibit at the Museum of Naval Aviation in Pensacola, Florida. (Courtesy of National Museum of Aviation.)

The wreck of a Douglas SBD Dauntless is on display at the bottom-of-Lake-Michigan exhibit at the National Museum of Naval Aviation in Pensacola, Florida. (Courtesy of National Museum of Naval Aviation.)

This model of the *USS Wolverine* was recently displayed at Merrill C. Meigs Field, Chicago, Illinois. (Courtesy of Chicago Maritime Society.)

BIBLIOGRAPHY

Alden, John D., Commander, U.S. Navy, "When Airpower Rode on Paddle Wheels," *U.S. Naval Institute Proceedings*, May 1961.

"Battle of Lake Michigan," *Air Classics*, Volume 4 Number 2, 1967.

Bonner, Kit, "The Great Lakes Paddle Wheel Carriers," *Sea Classics*, August 1995.

Chesneau, Roger, *Aircraft Carriers of the World: 1914 to the Present*, Naval Institute Press, 1984.

Clark, James M., "A Matter of Class: Part I," *Naval History*, September–October 1995.

"Coal Burning Carriers," *Naval Aviation News*, April 1954.

Czachor, Larry, "Secret Mission," *Chicago Tribune Magazine*, October 26, 1986.

Dalitsch, James Wm., "USS Wolverine (IX64) and USS Sable (IX81): The Great Lakes Aircraft Carriers," U.S. Coast Guard Academy paper, April 28, 1992.

Davis, Robert, "Chicago Maritime Museum Hopes to Float a Memory," *Chicago Tribune*, December 31, 1969.

Dornfeld, A.A., "Steamships after 1871," *Chicago History*, Spring 1977.

Gibbons, Jerry, "Lake Michigan Aircraft Carriers," *Bridgeview Independent*, January 25, 1990.

Greenwood, John Orville, M.B.A., *Namesakes II*, Freshwater Press, Inc., Cleveland, Ohio, 1978.

Grosvenor, Melville Bell, "The New Queen of the Seas," *National Geographic Magazine*, July 1942.

Hausner, Les, "At Last—A Happy Landing After Long Sleep in the Deep," *Chicago Sun-Times*, October 24, 1990.

Higgins, Brian, "Fresh Water Flat Tops," *Inland Seas*, #4, 1987.

"The Largest Sidewheeler in the World, Steamer 'Seeandbee,'" *Marine Review*, November 1912.

Laudermilk, John, "The *USS Wolverine*," Chicago Maritime Society, Fall 1991.

Lisagor, Peter, "Navy to Junk the Sable, Lake Aircraft Carrier," *Chicago Daily News*, June 28, 1948.

McDonell, Michael, "Great Lakes Carrier," *Naval Aviation News*, March 1972.

McKee, Jack E., "Carrier Training on Lake Michigan," *Military*, August 1994.

Mooney, James L. [ed.], *Dictionary of Naval Fighting Ships*, Naval Historical Center, 1981.

Nugent, CTC Edward E., USN (Ret), "The Paddle Wheelers of the 1940s," *Foundation*, Spring 1994.

"Rare World War II-Era Navy Plane to be Restored at Pensacola Museum," *Florida Times-Union*, Jacksonville, December 29, 1990.

Ratigan, William, *Great Lakes Shipwrecks and Survivals*, Galahad Books, 1960.

Ruth, Daniel, "Navy 'Guns' Roam the Lake," *Chicago Sun-Times*, November 9, 1988.

"The Side-Wheel Carriers," *American Heritage*, February/March 1987.

Steinberg, Neil, "Practicing Pilots Lost Warplanes Aplenty in the Lake," *Chicago Sun-Times*, October 28, 1990.

Stern, Robert, *U.S. Aircraft Carriers in Action, Part 1*, Squadron/Signal Publications, Carrollton, Texas, 1991.

Terry, Clifford, "How 'Top Guns of '43' Developed their Aim," *Chicago Tribune*, November 9, 1988.

Terzibaschitsch, Stefan, *Aircraft Carriers of the U.S. Navy Second Edition*, Naval Institute Press, 1989.

"Top Guns of '43' To Air on Channel 11," *Great Lakes Bulletin*, February 16, 1990.

"U.S.S. Sable," *Buffalo News*, June 1943.

Van der Linden, Rev. Peter J., *Great Lakes Ships We Remember II*, Freshwater Press, Inc., 1984.

"Vet, Plane Have 'Crash Reunion,'" *Florida Times–Union*, Jacksonville, December 8, 1990.

"Want to Buy a Carrier?" *Flying*, February 1946.

Watson, Leland "Trig", "The Lake Michigan Experiment," *The Hook*, Fall 1992.

Wendt, Gordon, "In the Service of her Country," *Sandusky Register*, December 29, 1991.

"Wolverine's Last Run," *Chicago Herald American*, November 29, 1947.

"WTTW Journal: Top Guns of '43," *Public Affairs on Eleven*, November 9, 1988.